AI MADE SIMPLE FOR PARENTS

Teaching Children and Older Students to Understand, Use, and Thrive with Artificial Intelligence

DR. ROSIE MILLIGAN

Printed and Bound in the United States of America

Published and Distributed By
Professional Publishing House LLC
Los Angeles, California
Telephone: 323-750-3592
Email: professionalpublishinghouse@yahoo.com
www.Professionalpublishinghouse.com

Cover design: TWA Solutions
First printing April 2026

ISBN: 979-8-994-1309-2-6

About the Author

Dr. Rosie Milligan is a nationally recognized author, publisher, business coach, and technology advocate whose life's work has been devoted to empowering individuals and families to build a better future. As C.E.O. of Professional Business Management/ Consulting Services and founder of My Tech Academy, she stands at the intersection of business acumen and community transformation.

Dr. Milligan holds a Ph.D. in Business Administration and is a seasoned Senior Estate Planner. She is sought after nationwide for her business and financial coaching services, and her motto—"Erase 'NO,' Step Over 'CAN'T,' and Move Forward with LIFE"—has become a rallying cry for hundreds she has mentored.

She has authored thirty-six books and published more than 450 books for other authors. Her titles, including *What You Need to Know Before You Start a Business, Getting Out of Debt Made Simple, Understanding Credit Made Simple,* and *Creating a Budget Made Simple* have helped readers across the country achieve financial clarity and personal empowerment. She is widely recognized as the largest and fastest-growing African American female publisher in the nation.

Dr. Milligan is also the host of a weekly live Zoom talk show, Express Yourself Hour, and the founder and director of Black Writers on Tour. Her decision to write *AI Made Simple for Parents* grew from a deep conviction: that every family—regardless of income, zip code, or educational background—deserves access to the knowledge that will shape the next generation's opportunities.

Contact her at:
323-750-3592 • drrosie@aol.com • Drrosie.com

Author's Note

AI is not coming—it is already here. It is not a fad. It is not a passing trend. Artificial Intelligence is the new literacy, and just like reading and writing, every child must learn it in order to thrive in today's world. The gap between those who understand AI and those who do not is already becoming one of the defining divides of our time—in education, in employment, and in economic power.

I wrote this book because I believe no child should be left behind. Not the child in the private school with every advantage, and not the child in the under-resourced community fighting against the odds. Every child deserves a guide, and every parent deserves the confidence to be that guide.

I strongly encourage every parent and grandparent to embrace AI fully and without fear. You do not need to be a technology expert. You do not need a college degree in computer science. You need exactly what you already have: love for your child and the willingness to learn something new alongside them.

Time is of the essence. AI is reshaping education, business, healthcare, and everyday life at a pace none of us have seen before. Schools and educators are still debating what to teach, how to teach it, who should be taught, and who should do the teaching. While those debates continue, one truth remains: the responsibility to prepare our children for life begins at home. It has always begun at home.

AI is not something to be feared. In this book, I will show you how to use AI in ways that support learning, creativity, and confidence for children, teens, and parents alike. I will give you plain-English explanations, real activities you can do today, and clear prompts that turn any AI tool into a learning partner for your family.

> ***"When parents are informed, children are empowered. Make AI a family affair."***

— Dr. Rosie Milligan

Table of Contents

CHAPTER 1

The Family Pledge
A Commitment That Changes Everything

Every meaningful journey begins with a decision. Before the lessons start, before the activities begin, before you open a single AI tool with your child, I want you to make a decision. I want you to commit.

This chapter contains two pledges. One is for you—the parent, grandparent, or guardian. The other is for your child. These are not mere formalities. They are declarations of intention. When you read them aloud, sign them, and display them in your home, you are telling your child: "Your future matters enough for me to show up. I will not let fear or ignorance stand in the way." That message is more powerful than any technology you will ever put in a child's hands.

Post these pledges on the refrigerator, in your child's room, or somewhere they will see them every day. Let them be a daily reminder of what you are building together.

The Parent Pledge

A Commitment to My Children's Future

I PLEDGE:

To prepare my child for the world they will inherit—not just the world I grew up in.

To understand that Artificial Intelligence is the new literacy, and that knowledge is a form of wealth that can never be taken away.

To commit to learning enough about AI to guide my child with wisdom, confidence, and discernment.

Not to allow fear, intimidation, or lack of technical experience to rob my child of opportunity. I will not leave my child behind in a digital world that rewards preparation and punishes ignorance.

To accept my responsibility as a parent, grandparent, or caregiver to take the lead—knowing that schools may change, systems may fail, and technology will continue to evolve.

To teach my child that AI is a tool, not a master; a resource, not a replacement for character, creativity, or critical thinking.

To make learning about AI a family affair—one that strengthens communication, builds confidence, and supports long-term economic stability.

Above all, to pass down more than possessions. I will pass down knowledge, readiness, and vision so that my child and future generations are equipped to build, protect, and sustain generational wealth.

Signed with intention and purpose,

Parent/Caregiver: ______________________________

Date: ____

The Child's AI Pledge

I PROMISE:

To learn and grow in a world filled with new ideas and technology.

To understand that Artificial Intelligence is a tool that can help me learn, create, and solve problems—and to use it wisely, honestly, and with good character.

To keep thinking for myself, asking questions, and using my imagination. I know that technology should help my mind grow—not replace it.

To respect my parents, grandparents, teachers, and mentors as they help guide me. I understand that learning is a gift that prepares me for my future.

To use what I learn to help myself, my family, and my community. I will work hard, stay curious, and prepare for the opportunities ahead.

I believe my future matters. I believe learning is power.
I believe I can build a strong and successful life.

Student: ________________________________

Age: ______________ Date: ______________________

CHAPTER 2

What Is AI? Plain Language for Every Parent

Before you can guide your child, you need to feel confident yourself. This chapter does one thing: it explains AI, Chat, and GPT in plain, everyday English, the kind of language we actually use at the kitchen table, not in a laboratory.

You do not need a technology background to understand this. If you can explain how a microwave works—press buttons, heat food, done—you can understand AI well enough to teach your child. Let's break it down.

What Is Artificial Intelligence (AI)?

Artificial Intelligence is simply a computer that has been taught how to think a little like a human. It can learn, remember, answer questions, solve problems, and help you get things done faster, but it only knows what it has been taught. AI does not have a mind of its own. It does not have feelings, opinions, or a hidden agenda. AI works for you.

Think of AI like a very smart assistant that never gets tired, never gets frustrated, and is always available. It does not judge you for asking a basic question. It does not get impatient when you ask the same thing twice. It simply does its job.

AI is already in your daily life right now, whether you realize it or not. When your phone's keyboard suggests the next word you might type—that is AI. When a streaming service recommends a movie you might enjoy based on what you've watched before—that is AI. When you speak to a voice assistant to set a timer or ask for directions—that is AI. It has been quietly working in the background for years. The only thing that changed is that now it can hold a full conversation with you.

> ***"AI is not a creature. AI is not a monster. AI is a tool—like a telephone, like a calculator, like a book. The question is never whether to have the tool. The question is whether you know how to use it."***

What Is "Chat"?

When we talk about "chat" in relation to AI, we simply mean having a conversation with the computer. Instead of pushing through complicated menus and buttons, you just type or speak what you want, and the AI talks back. It is as close to having a conversation with a knowledgeable assistant as technology has ever come.

You might type: "Explain photosynthesis to my 9-year-old in simple words." And the AI will do exactly that—immediately, clearly, and at exactly the level you asked for. That is the power of chat. You are in charge of the conversation. You set the direction. You ask the follow-up questions. The AI follows your lead.

What Is GPT?

GPT stands for Generative Pre-trained Transformer. Do not let those words intimidate you. Here is what they mean in plain English:

- Generative means it creates new responses, it does not just pull up a pre-written answer from a database.
- Pre-trained means it studied an enormous amount of text—books, articles, websites, conversations, before you ever talked to it.
- Transformer is the technical name for the type of computer design that makes it all work.

In everyday language: GPT is the brain behind the chat. It is why the computer can write, explain, answer, and talk in a way that sounds natural and human. It is why when you ask it a question, it doesn't give you a robotic, one-word answer—it gives you a thoughtful, organized response.

Putting It All Together

In Plain Language:

- AI is the helper.
- Chat is the conversation.
- GPT is the brain that makes the conversation smart.

Think of It Like:

- AI = Your assistant
- Chat = Your conversation
- GPT = The knowledge behind the answers

Bottom line: Artificial Intelligence is a smart tool you can talk to, ask questions of, write with, learn from, and use to make life and business easier—without being a tech expert. It is the great equalizer of our time, and your family now has access to it.

What AI Is and Is Not

AI IS:

✓ A learning tool

✓ A research assistant

✓ A homework helper

✓ A writing coach

✓ A creativity partner

✓ A patient explainer

✓ Available 24/7

AI IS NOT:

✗ A replacement for your child's thinking

✗ A babysitter

✗ Always 100% accurate

✗ A shortcut for avoiding work

✗ A substitute for parental guidance

✗ Something that controls your child's mind

✗ A danger when used with supervision

Keep that chart some place where you can see it. When doubt creeps in, when you wonder if AI is doing more harm than good, come back to it. AI is a tool. You are in charge of how it is used.

How to Start Using AI Today

You do not need to download anything complicated to get started. The most widely used AI chat tools are available right in your web browser. Here are three ways to access AI immediately:

- ChatGPT (chat.openai.com) – Free to use, available on any device with internet.
- Google Gemini (gemini.google.com) – Built into Google's ecosystem, easy to access.
- Microsoft Copilot (copilot.microsoft.com) – Free and integrated into many devices.

All three are free at the basic level. All three work on your phone, tablet, or computer. Pick one and try it today. You will be surprised how easy it is to get started.

A Word to Parents Who Feel Behind:

You are not behind. You are right on time. The fact that you are reading this book means you are already ahead of parents who are still waiting. The learning curve for AI is shorter than you think. Most parents who try it for the first time say the same thing: "I cannot believe how easy that was."

Give yourself the grace you give your children when they are learning something new. Take it one step at a time. Ask questions. Make mistakes. Try again. That is how learning works—at any age.

CHAPTER 3

Why AI Matters for Children and Older Students

AI is not just for adults, businesses, or tech people. AI matters deeply for our children—and for our older students too. And the sooner parents understand that, the better off our families and communities will be.

Many parents are afraid of AI. They worry it will make their children lazy, kill their creativity, or stop them from thinking for themselves. That fear is understandable, but it is misplaced. AI is not here to replace thinking. It is here to support learning. Just like calculators did not destroy math, and books did not destroy memory, AI will not destroy our children's minds. The real danger is not AI. The real danger is leaving our children unprepared for the world they are about to enter.

Children Learn Early and Fast

I firmly believe AI should be introduced as early as kindergarten. Research has already shown us something powerful: by the age of six, children have already developed many of the habits of mind—curiosity, confidence, persistence, and how they feel about learning—that will shape their entire lives. How they think, how they approach challenges, how confident they feel when exploring something new—all of that starts early.

Children today are growing up in a digital world whether we like it or not. Tablets, phones, smart TVs, and computers are already part of their environment from the time they can reach out and touch a screen. AI is simply the next layer of their digital world—and when introduced correctly, with a parent beside them, it becomes a tool, not a crutch.

Used properly and with guidance, AI can:

- Help children ask better, more precise questions
- Strengthen reading comprehension and vocabulary
- Support math learning step by step, at the child's own pace
- Encourage curiosity instead of frustration when a concept is unclear

- Provide patient, judgment-free explanations that build confidence

AI does not tell children what to think. It helps them learn how to think. That distinction is everything.

Parents Are Overwhelmed, and That Is Real

Today's parents are busy. Many are working one job, two jobs, or juggling life just to survive. When they come home, they are exhausted. They love their children deeply, but they do not always have the time, energy, or training to help with schoolwork the way they wish they could.

On top of that, students are taught differently now. The math your child brings home tonight may look nothing like the math you learned twenty years ago. Parents look at homework and say, "This is not how I learned it"—and they are right. That leaves parents feeling inadequate, frustrated, and sometimes embarrassed to admit they cannot help.

AI helps fill that gap. AI can explain math problems step by step using today's methods. AI can break down reading assignments in plain language. AI can help a child understand a lesson without making the child—or the parent—feel stupid or ashamed. That is not replacing parenting. That is supporting parenting. That is being resourceful.

When Resources Are Missing, AI Becomes the Bridge

Many under-resourced schools do not have enough books. That is a fact. Many families struggle with literacy. That is a fact. Libraries are underfunded, closing, or unsafe for children to visit alone. That is a fact. Tutors cost money most families simply do not have. That is a fact.

So, what happens when a child needs help, and the traditional support systems are gone? AI tools, used with parental guidance, can step into that gap. With proper supervision, AI can:

- Read stories aloud to children who need practice with reading
- Help struggling readers practice safely and privately at home
- Support homework when no tutor is available
- Give children access to explanations and information their schools may lack
- Provide enrichment that goes far beyond what underfunded classrooms can offer

> ***"AI does not judge. AI does not shame. AI does not get tired. And AI is available at two in the morning when your child suddenly remembers a project is due tomorrow."***

For families who have been locked out of educational resources for generations, AI is not just a convenience. It is an equalizer. It levels the playing field in a way that very few tools in history have been able to do. When a child in a rural community or an inner-city neighborhood has the same access to explanations, practice, and support as a child in a wealthy suburb—that is justice. That is opportunity. And it is available right now.

Older Students Need AI, Too

AI is not just for little children. Middle school students, high school students, college students, and adult learners all benefit from AI—especially those who fell behind, were never properly supported, or carry discouragement about learning from past experiences.

Many older students are not lazy. They are discouraged. They were told they were not smart. They were passed through a system that did not take the time to reach them. AI helps them re-enter learning without shame or fear.

Older students use AI to:

- Organize their thoughts and structure essays before they write
- Study for exams with customized quizzes and explanations
- Understand difficult subjects through multiple explanations until one finally clicks
- Build confidence in learning again, one small success at a time
- Prepare for job applications, interviews, and professional communication

Creativity Is Not Lost—It Is Expanded

One of the most common fears parents express is this: "If AI does the work, my child won't learn to be creative." I want to address that fear directly.

AI does not destroy creativity. Lack of opportunity destroys creativity. Shame destroys creativity. Being told you are not smart and watching others get ahead while you struggle—that destroys creativity. When children have support and understanding, they become more confident. When they are confident, they create. When they create, they innovate. When they innovate, they lead.

AI can help a child:

- Write stories they never could have organized on their own
- Explore ideas that started as a tiny spark of curiosity
- Ask "what if" questions and get real, thoughtful responses that fuel further thinking
- Build projects, presentations, and creative works they never believed were within their reach

Creativity does not disappear when children are supported. It blossoms. AI is not the enemy of the imagination. It is fuel for it.

The Parent's Role Is Unchanged

Let me say this clearly: AI does not replace parents. AI does not replace teachers. AI does not replace guidance, discipline, or values. Parents still teach right from wrong. Parents still model integrity, responsibility, and how to treat other people. Parents still shape the character that determines what a child does with every tool they are given.

AI is a learning assistant, not a parent. Parents who embrace AI are not giving up control. They are giving their children a fighting chance in a world that is already being shaped by this technology. The choice is not between AI and no AI. The choice is between AI with a wise parent's guidance, or AI with no guidance at all. One of those choices leads to power. The other leads to vulnerability.

Dear Parent: Please Stop the Fear

Fear will not protect our children. Education will.

We cannot protect our children by hiding technology from them. We protect them by teaching them how to use it wisely, ethically, and purposefully.

AI is already here. The question is not if your child will encounter it. The question is whether they will be prepared when they do—or whether they will be left behind, manipulated by it, or outcompeted because of it.

Parents who embrace AI are not reckless. They are forward-thinking. They are protecting their children's future. And in this fast-paced world, preparing your children is real love.

CHAPTER 4

A Brief History of AI
How We Got Here

Artificial Intelligence did not arrive overnight, and it did not arrive by accident. It grew from decades of questions, failures, breakthroughs, and persistence. Understanding where AI came from helps us understand what it is, what it can do, and why we should not be afraid of it.

The Story of AI

In the early 1950s, a British mathematician named Alan Turing asked a simple but radical question: "Can machines think?" He did not mean think with emotions or consciousness—he meant think well enough to solve problems, recognize patterns, learn from experience, and communicate in human language. He even created a test—now called the Turing Test—to see if a machine could hold a conversation convincing enough that a person could not tell they were talking to a computer.

That question cracked a door that could never be closed again. It invited scientists, mathematicians, and engineers around the world to imagine a new kind of machine—one that did not just calculate, but learned.

1956: AI Gets Its Name

In 1956, a group of researchers gathered at Dartmouth College for a summer workshop and officially named this field of study Artificial Intelligence. Among them were some of the greatest scientific minds of the 20th century. Their goal was clear: teach computers to do things that human minds do—understand language, solve logical problems, learn from experience, and make decisions.

In those early years, AI was used primarily by universities and the military to solve math problems, decode encrypted messages, and play strategy games like chess. These were important milestones, but they were also limited. The computers of that era were slow, expensive, and could only do exactly what they were explicitly programmed to do. Nothing more.

The 1960s–1970s: Big Hopes, Hard Lessons

The 1960s brought enormous excitement and enormous disappointment. Researchers predicted that thinking machines were just around the corner. They were wrong. Computers of that era did not have enough processing power, enough memory, or enough data to do the things AI theorists imagined. Programs could play chess, but they could not hold a conversation. They could solve equations, but they could not understand a sentence.

Funding dried up. Public confidence faded. Historians of technology call this period the first "AI Winter"—a long, cold stretch where progress slowed and interest waned. But the researchers who believed kept working. They kept asking questions. They kept building.

The 1980s: AI Becomes Practical

In the 1980s, AI came back to life in a different form. Instead of trying to build machines that thought like humans, researchers focused on something smaller and more achievable: expert systems. These were programs that copied how human experts in specific fields made decisions.

Doctors used them to help diagnose illnesses. Engineers used them to identify equipment failures.

Banks used them to evaluate loan applications. These systems were not brilliant—they were useful. And useful was enough to reignite interest and funding.

The 1990s–2000s: Machines Begin to Learn

This is where AI changed fundamentally. Instead of programming a computer with every possible rule and instruction, researchers discovered something revolutionary: you could teach a computer to learn from examples. Feed it enough data, and it would figure out the patterns on its own. This is what we now call machine learning, and it changed everything.

Machine learning allowed AI to be trained on massive amounts of real-world data—credit records, medical images, customer behavior, spoken language—and to improve its accuracy over time. AI started being used for credit scoring, fraud detection, search engines, and speech recognition. In 1997, an AI called Deep Blue defeated world chess champion Garry Kasparov, a moment that shook the world and announced that machine intelligence was no longer science fiction.

2010s to Today: AI for Everyday People

With faster computers, billions of smartphones, and the internet connecting the world into a single information ecosystem, AI moved from laboratories and boardrooms into the palms of our hands. The 2010s brought voice assistants, recommendation engines, real-time translation, and self-driving car research. By the early 2020s, AI could write, draw, compose music, analyze medical images, and hold sophisticated conversations.

In 2022, tools like ChatGPT became available to the general public and crossed fifty million users in just five days, the fastest adoption of any technology in history. Overnight, AI went from something that belonged to corporations and governments to something that a parent in any neighborhood could use to help their child with homework at no cost.

What changed is access. What once belonged only to scientists and large institutions is now available to everyone: seniors, parents, small business owners, and families building a legacy. AI didn't just arrive—it evolved. And now it's your turn to use it wisely.

A Note for Parents to Share with Children:

AI has been growing and learning for almost seventy years. The people who built it were curious, patient, and persistent. They failed many times before they succeeded.

That is exactly what we want our children to be: curious, patient, and persistent. Not afraid of hard problems. Not defeated by failure. Just like the scientists who made AI possible.

CHAPTER 5

AI Activities for Young Children Ages 3–6

Ages three to six are some of the most important years in a child's development. This is the window when confidence, curiosity, fearlessness, and the love of learning are formed. A child who grows up comfortable with technology, who sees it as a fun tool rather than something scary or forbidden, will carry that confidence for life.

At this age, AI activities are not about screens and typing. They are about talking, playing, imagining, and guiding. Think of AI as a learning buddy—something you and your child explore together. Every activity in this chapter is parent-led. You are always in charge. The AI responds to you. You respond to your child. That is the model.

You do not need special equipment. A smartphone, a tablet, or a computer with any free AI tool will work perfectly. Let's get started.

Activity 1: Talk to the Smart Helper

Purpose:
Introduce your child to the concept of asking questions and getting answers from a digital assistant. Build listening skills and vocabulary while making technology feel friendly and safe.

Try These Prompts:
→ *Tell us a story about a happy dog named Max.*
→ *What color is the sky, and why is it that color?*
→ *Can you help us count to twenty?*
→ *What sound does an elephant make?*

What Your Child Learns:
✓ How to ask clear, specific questions
✓ Active listening and comprehension
✓ Vocabulary development
✓ That technology responds to spoken words—and to us

Parent Tip: *After each AI response, ask your child: "What did you hear?" or "What did you learn?" This teaches them to listen for information, not just noise.*

Activity 2: Build-A-Story Together

Purpose:

Use AI to co-create a story with your child, keeping the child's imagination at the center of the creative process. This teaches that AI is a tool that responds to our ideas—not a machine that replaces them.

Try These Prompts:

→ *Create a short bedtime story for a 4-year-old about a butterfly who learns to share.*

→ *Now make the butterfly meet a friendly caterpillar.*

→ *Give the butterfly a name that my child chooses.*

→ *Make the story have a happy ending.*

What Your Child Learns:

✓ Story structure and narrative flow

✓ Emotional intelligence and empathy

✓ Cause and effect thinking

✓ That creativity comes from the person—AI just helps organize it

Parent Tip: *Pause after every few sentences and ask: "What do you think should happen next?" The child is the author. AI is just the pen.*

Activity 3: Draw What the Helper Describes

Purpose:
Develop listening comprehension and imagination by having your child draw what the AI describes. This keeps the child's mind and hands active, and shows that AI gives ideas while humans give them life.

Try These Prompts:
→ *Describe a silly, colorful monster for a 5-year-old to draw.*
→ *Describe a magical house in a forest.*
→ *Tell us about a friendly dragon with stripes.*

What Your Child Learns:
✓ Listening comprehension
✓ Creative visualization
✓ Fine motor skills through drawing
✓ The idea that AI describes but humans create

Parent Tip: *Frame it as a game: "Let's see if your picture matches what the helper described!" No wrong answers. All drawings are a win.*

Activity 4: Count and Sort with the Smart Voice

Purpose:
Use AI as a patient math tutor for early number skills. Lay out physical objects and involve the AI as a guide.

Try These Prompts:
→ *Help us count these 5 blocks and 3 toy cars. How many are there in all?*
→ *We have 8 things. If I take away 3, how many are left?*
→ *Can you give us a counting song?*

What Your Child Learns:
✓ Early number concepts and counting
✓ Basic addition and subtraction
✓ Confidence in math as something friendly and achievable

Parent Tip: *Always count with your hands and objects first. Use AI to check and confirm. The physical touch matters at this age.*

Activity 5: Good Choice/Safe Choice Game

Purpose:
Use AI to introduce character development and safety concepts in a non-threatening, game-like format.

Try These Prompts:
→ *Give an example of a good choice a 5-year-old could make when they are angry.*
→ *What should a child do if a stranger tries to talk to them when no parent is nearby?*
→ *What is a kind thing a child can do for a friend who is sad?*

What Your Child Learns:
✓ Decision-making and self-regulation
✓ Safety awareness
✓ Empathy and social skills
✓ That grown-ups have answers and AI can help explain them

Parent Tip: *After every AI answer, ask your child: "Do you think that was a good choice? Why?" The conversation is more important than the answer.*

The Parent Is Always the Boss

Say this out loud, often, and cheerfully: "AI is a tool. Mommy/Daddy/Grandma is the boss."

This simple phrase does three important things. It sets healthy boundaries around technology. It reinforces your authority as a parent. And it removes fear from your child's relationship with AI by making the hierarchy clear and safe.

What These Activities Teach Parents:

✓ AI is not dangerous when you are present and in charge.

✓ You do not need technical skills to use AI with your child.

✓ AI can strengthen your bond with your child rather than compete with it.

✓ Early positive exposure to technology removes fear later in life.

✓ A parent's presence is the most important ingredient in every activity.

CHAPTER 6

AI Activities for Tweens and Teens Ages 7–18

As children move into the middle school years and beyond, their relationship with AI needs to grow with them. The stakes are higher. The temptations are real. Teens who are not guided on how to use AI appropriately will either misuse it, using it to cheat, to avoid thinking, or to access harmful content or they will be intimidated by it and fall behind their peers who are using it effectively.

The activities in this chapter are designed to make AI a tool for growth, not a shortcut. They are built around real skills that will matter for the rest of your teen's life: critical thinking, financial literacy, decision-making, writing, career planning, and character. Each activity includes a family component, because the most powerful learning happens in relationship.

Activity 1: AI as a Homework Helper, not a Cheater

Purpose:
Teach the difference between using AI as a tutor and using it as a ghostwriter. This is one of the most important lessons a teen can learn.

Try These Prompts:
→ *Explain this math concept step-by-step as if I am in 8th grade.*
→ *I do not understand this paragraph. Can you break it down for me?*
→ *Quiz me on this topic with five questions to see if I really understand it.*
→ *Explain why this answer is correct, not just what the answer is.*

What Your Child Learns:
✓ How to use AI to understand—not just to complete
✓ Study strategies and self-testing
✓ Academic integrity
✓ The difference between learning and cheating

Parent Tip: *The test is simple: if your teen can explain what AI helped them learn in their own words, they learned it. If they cannot, they only copied. Ask them to explain it back to you after every session.*

Activity 2: AI Career Vision Board

Purpose:
Connect AI directly to your teen's future income and career potential. Make the connection real and personal.

Try These Prompts:
→ *What careers that pay well also use AI skills?*
→ *What should a teenager start learning now to prepare for a career in healthcare? In business? In technology? In the arts?*
→ *What careers will still exist in twenty years, and why?*
→ *What is the average salary for someone who understands AI in my field of interest?*

What Your Child Learns:
✓ Career awareness and long-term planning
✓ Connection between education and economic opportunity
✓ The role AI plays in nearly every industry
✓ That preparation now creates options later

Parent Tip: *After the activity, ask your teen: "Which career excited you most? Which one surprised you?" Then follow the conversation wherever it goes.*

Activity 3: Fact-Checker Challenge: Truth vs. Trash

Purpose:
In a world flooded with misinformation and social media manipulation, the ability to verify information is one of the most powerful skills a young person can have.

Try These Prompts:
→ *Is this claim true? Please show me facts from reliable sources.*
→ *What are signs that a news story might not be accurate?*
→ *Who might benefit from spreading false information about this topic?*
→ *Where should I go to verify this claim?*

What Your Child Learns:
✓ Critical thinking and media literacy
✓ How to distinguish credible sources from unreliable ones
✓ Understanding that information can be weaponized
✓ Epistemic confidence—knowing how to find the truth

Parent Tip: *Find a viral post or controversial headline together and run it through this process. Make it a habit, not a one-time lesson.*

Activity 4: AI-Assisted Decision-Making

Purpose:
Use AI to practice thinking through real-life scenarios before they happen. Prevention is always more effective than repair.

Try These Prompts:
→ *Give me three smart choices and three risky choices in this situation: [describe scenario].*
→ *What are the possible long-term consequences of this decision?*
→ *How might I feel about this choice in five years?*
→ *What would a wise mentor tell me to do here?*

What Your Child Learns:
✓ Consequences thinking and delayed gratification
✓ How to slow down before reacting
✓ Identifying peer pressure and manipulation
✓ Personal accountability and self-awareness

Parent Tip: *Use real situations your teen is facing. The more relevant the scenario, the more powerful the learning. This is not hypothetical—this is practice for real life.*

Activity 5: AI as a Writing and Voice Coach

Purpose:
Build communication skills that will matter in school, in careers, and in life. Help your teen find their voice—with AI as the coach, not the ghostwriter.

Try These Prompts:
→ *Help me rewrite this paragraph so it sounds clearer and more confident, but still sounds like me.*
→ *What are the strongest parts of what I wrote? What can be improved?*
→ *Help me organize my ideas before I start writing.*
→ *Give me feedback, not the answer.*

What Your Child Learns:
✓ Writing clarity and structure
✓ Self-editing and revision
✓ Communication confidence
✓ That their voice has value and is worth refining

Parent Tip: *Have your teen read both versions—original and revised—aloud. Ask: "Which one sounds more like the person you want to be?"*

Activity 6: Budgeting and Money Sense with AI

Purpose:
Introduce financial literacy in a practical, hands-on way. Money skills are life skills, and teens are ready to begin building them.

Try These Prompts:
→ *Create a simple monthly budget for a teen earning $300 a month.*
→ *What percentage of my income should I save vs. spend vs. give?*
→ *What are the worst money habits teens develop, and why are they hard to break?*
→ *Explain compound interest as if I am 15 years old.*

What Your Child Learns:
✓ Budgeting and spending discipline
✓ The difference between needs and wants
✓ Why saving early creates long-term wealth
✓ That financial habits formed now follow you into adulthood

Parent Tip: *Turn this into a family conversation about real money. Share your own financial lessons—the wins and the mistakes. Real stories teach more than any activity.*

Activity 7: AI Ethics and Character Conversation

Purpose:
Ground all AI use in values. Technology is only as good as the character of the person using it.

Try These Prompts:
→ *Should AI replace teachers? What are the arguments on both sides?*
→ *What are the dangers of using AI irresponsibly?*
→ *Should AI ever decide who gets a job, who goes to jail, or who gets a loan?*
→ *If you could set rules for how AI should be used in schools, what would they be?*

What Your Child Learns:
✓ Ethical reasoning and values clarification
✓ Understanding power and its potential for misuse
✓ That technology must be governed by character
✓ Critical perspective on the tools they use

Parent Tip: *This activity has no wrong answers. The goal is conversation, not conclusion. Ask: "Why do you feel that way?" as often as you can.*

CHAPTER 7

AI Prompt Cards for Teens
Think Smart. Choose Wise. Build Your Future.

The following prompt cards are designed for your teen to use independently, or with you. They are organized by life skill, not by subject. Each card tells your teen when to use it, what to ask, and what to remember. Print them, save them, or post them somewhere your teen will see them.

Before every session, remind your teen of the two rules that never change:

- AI is here to help you think better, learn faster, and plan smarter.
- AI is not here to replace your thinking, do your work, or make choices for you.

PROMPT 1: AI FOR SCHOOL HELP (WITHOUT CHEATING)

Use this when you are stuck, confused, or need a concept explained in a different way.

"Explain this lesson like I am in middle school/high school."
"Break this problem into simple steps I can follow."
"Help me understand why this answer is correct."
"Quiz me to see if I really understand this topic."
"I still do not understand. Can you explain it a different way?"

Remember: Learning means understanding. Copying means cheating. If you can explain it back to someone else, you learned it.

PROMPT 2: AI FOR SMART DECISION-MAKING

Use this before you react to a situation, not after. AI can help you slow down and think ahead.
"What could happen if I make this choice?"
"Give me three smart choices and three risky choices for this situation."

"Which choice protects my future?"
"What would I regret about this decision in five years?"
"What would a wise mentor tell me to do?"

Remember: A quick decision can cause a long problem. Slow down. Think first.

PROMPT 3: AI FOR ONLINE SAFETY

Use this when something online feels off. Trust your instincts—and then check.

"Is this type of message safe or suspicious?"
"How do scammers typically target teenagers online?"
"What should I do if a stranger contacts me online and asks personal questions?"
"How can I protect my personal information and privacy?"
"Is this too good to be true?"

Remember: If anyone online asks you to keep secrets, send photos, or click strange links—Stop. Tell a trusted adult. No exceptions.

PROMPT 4: AI FOR MONEY AND BUDGETING

Use this to build financial habits that your future self will thank you for.

"Create a simple budget for a teen earning $300 a month."
"How much should I save versus spend?"
"What are the worst money habits teenagers develop?"
"Explain how compound interest works in simple language."
"How much would I have at 30 if I saved $50 a month starting now?"

Remember: Do I need this or want this? Will Future-Me thank Present-Me for this choice?

PROMPT 5: AI FOR CAREER AND PURPOSE

Use this to think beyond right now—toward who you are becoming.

"What careers use AI in music, sports, business, or healthcare?"
"What skills should teenagers start developing now for high-demand careers?"

"What jobs will still exist in 10 to 20 years, and why?"
"What careers pay well and also help people?"
"What can I do this year to prepare for the career I want?"

Remember: Your future is built one skill at a time. Every skill you learn today is a door that opens later.

PROMPT 6: AI FOR WRITING AND CONFIDENCE

Use this to improve your communication skills. Your voice matters. Let AI help you sharpen it.

"Help me rewrite this to sound clearer and more confident—but still like me."
"What are the strongest parts of what I wrote?"
"Help me organize my thoughts before I start writing."
"Give me feedback on my argument, not just my grammar."
"Help me write this email / message in a professional way."

Remember: AI helps shape your words, not replace your voice. You still have to think. You still have to feel. AI just helps you express it better.

PROMPT 7: AI AND CHARACTER CHECK

Use this when you need to think through the ethics of a situation, not just the logistics.

"What are the dangers of misusing AI?"
"How can AI be used to help people—and also to harm people?"
"Should AI be used in this situation? Why or why not?"
"What does responsible use of this technology look like?"
"Am I using this tool to grow, or to avoid growing?"

Remember: Just because you can does not mean you should. Does this choice reflect who you want to be?

CHAPTER 8

Keeping Children Safe in an AI World

Technology without safety education is like giving a child car keys without teaching them to drive. AI opens incredible doors, but some of those doors lead to dangers that parents must be aware of and prepared for. This chapter is not here to frighten you. It is here to equip you.

The goal is not to ban AI or restrict your child from the internet. The goal is to raise a child who knows the difference between a safe space and a dangerous one—online and in the real world.

The Risks Every Parent Should Know

AI makes certain kinds of harm easier than they have ever been before. Here is what parents need to be aware of:

Misinformation and Deepfakes

AI can create fake images, fake videos, and fake audio that look and sound completely real. Your child may see a video of a famous person saying something they never said. They may receive a voice message that sounds like a family member but is not. Teaching children to verify information before believing or sharing it is no longer optional, it is essential.

Online Predators Using AI

Predators are increasingly using AI to craft more convincing, personalized messages to approach young people. The days of poorly written, obvious scam messages are fading. Today's messages can be warm, specific, and emotionally intelligent—because AI helps write them. Teach your child: no secret is worth keeping if it involves an adult online. No legitimate relationship requires secrecy from parents.

Academic Dishonesty

The temptation to use AI to write an essay, complete an assignment, or take a test is real—and it is growing. Schools are developing detection tools, but more importantly, students who cheat rob themselves of the learning they need. The skills AI is supposed to support are the skills they will need for the rest of their lives.

Teach your child that integrity is not something they practice only when someone is watching.

AI Addiction and Overdependence

Just like social media and video games, AI tools can become habits that replace thinking rather than support it. If a child reaches for AI every time a problem feels hard—before they have tried it themselves—they are building a dependence that will weaken them. The rule in your house should be simple: try it yourself first. Use AI to understand, not to avoid.

Rules for AI Use in Your Home

Suggested Family AI Guidelines:

1. AI tools are used in shared spaces, not behind closed doors.
2. Parents have the right to review what their child has used AI for.
3. AI may never be used to complete school assignments that are meant to be done by the student.
4. No personal information—address, phone number, school name, photos—is shared with any AI tool.
5. If something AI says feels wrong, scary, or confusing, come to a parent immediately.

6. AI is for learning, creating, and solving problems. It is not a babysitter, a therapist, or a secret keeper.
7. Time limits apply to AI use just as they do to video games and social media.

Post these rules in your home alongside the pledges from Chapter 1. Rules without relationship become rebellion. But rules within a relationship of trust become values.

How to Talk to Your Child About AI Safety

You do not need a script. You need a conversation. Here are some questions to ask your child regularly:

- "What did you use AI for today? Can you show me?"
- "Did anything AI said confuse you or bother you?"
- "Is anyone online trying to talk to you in a way that feels strange?"
- "Do you know the difference between what AI tells you and what is actually true?"

These conversations do not need to be long. They need to be consistent. A family that talks openly about technology is a family that is protected.

CHAPTER 9

AI and Economic Opportunity Building Generational Wealth

I have spent my entire career helping individuals and families build financial stability. I have written books on getting out of debt, understanding credit, and creating budgets. And in all my years of work, I have never seen a tool with more potential to democratize economic opportunity than AI.

This chapter is about money. It is about the connection between AI literacy and economic power. It is about what happens when our children are equipped—and what is lost when they are not.

The AI Economy Is Already Here

Companies across every industry: healthcare, finance, education, retail, entertainment, manufacturing, real estate are rapidly integrating AI into their operations.

They are not waiting to see how it turns out. They are hiring people who understand AI and letting go of people who cannot adapt.

According to multiple workforce studies, AI literacy is quickly becoming a required skill across job categories, not just in technology fields but in other fields such as:

- A customer service representative who can use AI tools effectively.
- A teacher who can personalize learning with AI.
- A small business owner who can automate operations with AI.
- A healthcare worker who can interpret AI-generated data.

These are not future scenarios. They are happening right now.

The families who understand this,who prepare their children for this reality—are the families who will move forward. The families who wait are the families who will be left behind.

AI Is a Leveling Tool

For generations, access to quality education, professional networks, and financial resources has not been equally distributed. Families with wealth had access to private tutors, prep schools, enrichment programs, and professional mentors. Families without wealth did not.

AI does not eliminate all of those inequalities. But it does something historic: it puts a world-class, patient, personalized tutor, business advisor, writing coach, and research assistant in the hands of any child with a device and an internet connection. That is not a small thing. That is a generational shift.

The child who grows up knowing how to use AI well—to research, to write, to problem-solve, to communicate, to build businesses—has a significant advantage that did not exist twenty years ago. Our job as parents is to make sure that advantage belongs to our children.

How Families Can Use AI for Financial Growth

AI is not just for children and students. It is for you, too. Here is how parents and families can begin using AI to build financial strength:

- Business planning: Ask AI to help you write a business plan, identify your target market, or research your competition. A business idea that might have taken months to evaluate can now be explored in hours.
- Financial literacy: Ask AI to explain complex financial terms—compound interest, credit scores, investment accounts, life insurance—in plain language. Then teach what you learn to your children.

- Resume and job applications: Use AI to help improve a resume, prepare for a job interview, or write a professional cover letter. These are skills that translate directly to higher earnings.
- Real estate education: Ask AI to explain how mortgages work, what equity means, and how to evaluate a property. Real estate has been one of the most reliable wealth-building tools in American history.
- Small business operations: AI can write emails, create marketing copy, suggest pricing strategies, and help organize business plans for entrepreneurs and small business owners.

> ***"Knowledge is the only form of wealth that cannot be taxed, stolen, or taken away. When you learn something, you own it forever. When you teach your children, they own it forever."***

A Note on Generational Wealth

Generational wealth is not only money. It is habits. It is mindset. It is the knowledge that gets passed from parent to child, from grandparent to grandchild. When you sit down with your child and explore AI together, you are passing down something more valuable than any possession: the understanding that learning is power, that new tools are opportunities rather than threats, and that your family does not wait to be included—it prepares to lead.

That is the real legacy this book is asking you to build.

CHAPTER 10

A Final Word to Every Parent

To every parent, grandparent, and guardian who has read this book:

I want to start by saying something that does not get said enough: Thank you. Thank you for picking up this book. Thank you for caring enough about your child's future to learn something new. Thank you for not letting fear, busyness, or overwhelm have the final word. That took something. And I want you to know it matters.

You Do Not Have to Know Everything to Lead

I understand how uncomfortable it feels to try to guide a child into territory you have not fully explored yourself. I understand the self-doubt that comes with sitting down at a computer and not being sure what you are doing. I understand the weight of feeling behind in a world that moves so fast.

But here is what I know about parents: you have been figuring things out your whole lives. You figured out how to love someone more than yourself. You figured out how to sacrifice sleep, comfort, and personal plans for the sake of a child. You figured out how to hold a family together during hard times. You did not have a manual for any of that either—but you did it anyway.

This is the same thing. It is new territory, but you are not new to rising to a challenge. Give yourself the same grace you give your children when they are learning something hard.

Make It a Family Affair

Do not try to learn AI alone and then present it to your children as finished knowledge. Learn with them. Explore with them. Be wrong with them. Laugh with them. Let them see you make mistakes and keep going. That models something more important than any lesson in this book: it shows them that adults do not stop learning, that intelligence is not fixed, and that curiosity is not something you outgrow.

Get the entire family involved. Let grandparents ask questions, and let the grandchildren explain the answers. Let teenagers show younger siblings what they discover. Let parents and children teach each other. That kind of learning sticks in a way that no worksheet or assignment ever will.

Schedule Real Time

Do not wait for the perfect moment. Do not wait until you feel ready. Do not wait until life slows down because it will not. Schedule time, actual, marked-on-the-calendar time to do one activity from this book with your child. Just one. This week. Put it on the calendar the same way you put doctor appointments and school events. Your child's future deserves that level of commitment.

And then do it again the next week. And the week after that. Progress is not made in great leaps. It is made in consistent, small steps. One activity. One conversation. One discovery at a time.

What You Will Find Along the Way

You will be surprised at what you discover—not just about artificial intelligence, but about your child. You will learn how they think. You will see their curiosity light up. You will discover strengths you did not know they had, and you will find places where they need support that you can now provide. You will also discover something about yourself: that you are more capable than you thought, more adaptable than you feared, and more important to your child's future than any technology will ever be.

> ***"AI may change the world, but character, guidance, and family will always shape it."***

Do Not Wait. The Future Is Not Waiting.

AI is moving fast. The world is changing quickly. The window of opportunity to give your child a foundation in this technology is open right now, but windows do not stay open forever. The families who act today will be ahead of the curve in five years. The families who wait will be catching up.

Be intentional. Be forward-thinking. Be the parent who said: "I may not know everything about this, but I will not let that stop me from giving my child the best chance I can."

That is the kind of parent this world needs. That is the kind of parent your child needs. And based on the fact that you read this book to the very last page, I believe that is exactly the kind of parent you are.

QUICK REFERENCE

AI Starter Prompts for Every Parent

Use these prompts today. No setup required. Just open any free AI tool and type.

For Helping with Homework

- "Explain [topic] in simple language for a [grade] student."
- "Break this math problem into steps my child can follow."
- "Quiz my child on [subject] with five questions."

For Family Learning

- "Tell us a short story with a lesson about [value: honesty, kindness, courage]."
- "Explain how [household item] works in a fun way for a young child."
- "Give us a fun trivia game about [topic]."

For Your Own Learning

- "Explain [financial term] in plain English."
- "Help me write a professional email to [employer, landlord, school]."
- "What are practical steps to start a small business in [your field]?"

For Teens

- "Help me understand this concept—explain it three different ways."
- "What careers in [field] are growing because of AI?"
- "Create a study plan for my exam next week."

www.ingramcontent.com/pod-product-compliance
Lightning Source LLC
LaVergne TN
LVHW090430210926
848480LV00002B/768
9798994130926